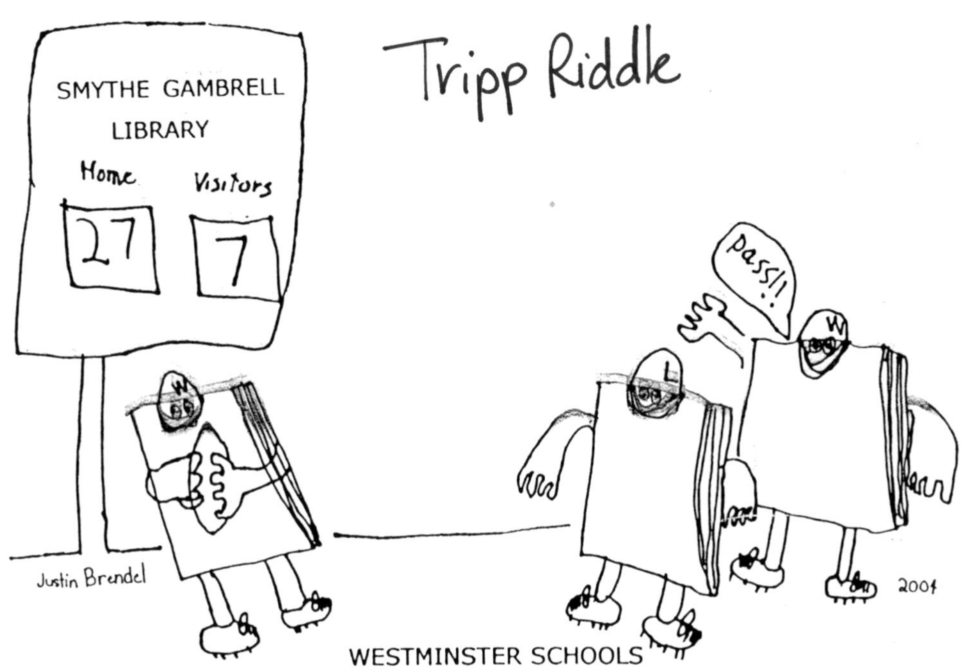
PRESENTED BY
Tripp Riddle
SMYTHE GAMBRELL
LIBRARY
Home
Visitors
27
7
pass!!
Justin Brendel
2004
WESTMINSTER SCHOOLS

THE LIBRARY OF SATELLITES

COMMUNICATIONS SATELLITES

ANN BYERS

the rosen publishing group's
rosen central

Published in 2003 by The Rosen Publishing Group, Inc.
29 East 21st Street, New York, NY 10010

First Edition

Library of Congress Cataloging-in-Publication Data

Byers, Ann.
Communications satellites / Ann Byers.
v. cm. — (The library of satellites)
Includes bibliographical references and index.
Contents: How satellites work—The "global village"—The business of satellite communication—What comes next?.
ISBN 0-8239-3851-4 (library binding)
1. Artificial satellites in telecommunication—Juvenile literature.
[1. Artificial satellites in telecommunication.] I. Title. II. Series.
TK5104 .B95 2003
384.5'1—dc21

2002007527

Manufactured in the United States of America

TABLE OF CONTENTS

INTRODUCTION

On July 10, 1962, the eyes of three nations—England, France, and the United States—were fixed on hundreds of thousands of television screens. People in the United States and Europe watched intently, eagerly awaiting pictures beamed from across the thousands of miles of ocean that separated them. This day was to be the first live broadcast of sight and sound across the Atlantic.

Before this day, news between Europe and America had to travel by telegraph or telephone, carried long distances over wires or cables. Any televised pictures had to be recorded on tape, sent across the ocean by plane, and played back hours after the event occurred. The promise of July 10, 1962, was one of instant communication: People in England and France would see what was happening in the United States the very moment it was occurring. This represented nothing less than a communications revolution, and excitement was high.

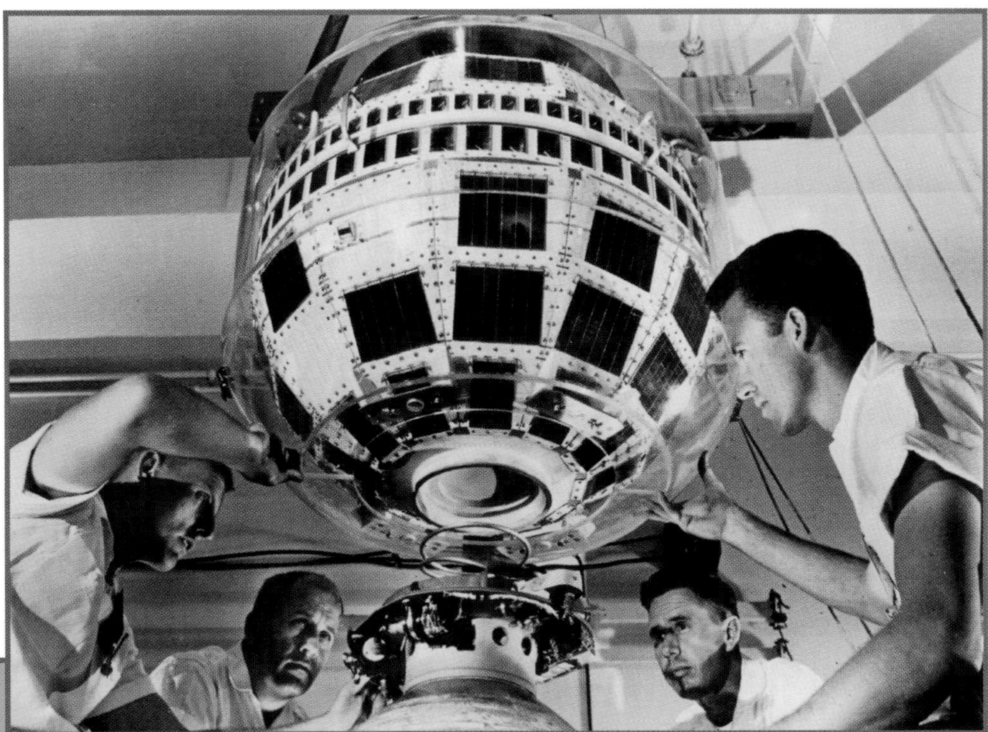

Technicians connect the *Telstar* satellite to the Delta rocket before its launch into orbit from Cape Canaveral, Florida, on July 10, 1962. The successful launch enabled the first live television broadcasts to be seen simultaneously in England, France, and the United States.

TELSTAR IS BORN, AND A NEW ERA BEGINS

At no place was excitement higher than Bell Telephone Laboratories. Scientists at Bell Labs, the research and development arm of American Telephone and Telegraph (AT&T), had made this technological miracle possible. For almost two full years, they had worked on designing and building a communications satellite that could be placed high enough in the sky to receive signals from one location and beam them down to another, thousands of miles away. One-third of all the employees at the laboratories had been dedicated to the project. They called their experimental satellite *Telstar*.

AT&T had the technology, the manpower, and the money to build *Telstar* in its Bell Labs. But to provide international communication, the company needed earth stations in more than one country. Earth stations are the places that both send and receive signals from a satellite. For example, a television signal sent from a U.S. earth station is received by a satellite, then relayed to an earth station in Europe. The signal is then processed and broadcast on European television. This entire process is virtually instantaneous.

A few years before *Telstar* was built, AT&T executives had worked with the government-operated telephone organizations of England and France to lay telephone cables across the Atlantic Ocean so that citizens of the three countries could communicate with each other quickly, easily, and clearly. Now these two European nations were eager to help in this exciting new broadcast satellite venture. Both countries had built, at their own expense, earth stations that could send and receive signals to and from the American satellite. AT&T had constructed its own ground station in Andover, Maine.

Getting the *Telstar* satellite into orbit had been a huge challenge. Since this was a private venture rather than part of a government program, AT&T had to find a way to get the satellite into space and pay for the launch itself. Finally, the National Aeronautics and Space Administration (NASA), a U.S. government agency devoted to the exploration and study of space and Earth's atmosphere, agreed to take *Telstar* more than 500 miles (804.7 km) into the sky aboard a Thor Delta rocket for the price of $3 million. The Delta could carry a maximum load of 180 pounds (81.6 kilograms), and *Telstar* weighed 171 pounds (77.6 kilograms). By the summer of 1962, everything was in place for the launch of the world's first communications satellite.

On July 10, reporters and dignitaries gathered at Cape Canaveral in Florida to watch the launch. Technicians met at Andover, Maine, at Pleumeur-Bodou, France, and at Goonhilly Downs, England, to aim their antennas at *Telstar*, which looked like little more than a metal ball almost three feet (0.9 meters) around. Ordinary citizens in all three countries gathered around their television sets, eager to see the first live video transmission beamed from space.

Late in the afternoon, when *Telstar* was on its sixth orbit, it switched on. In three countries, people held their breath. Within seconds, on thousands of television screens, a picture of white clouds in a blue sky gave way to the red, white, and blue of the American flag waving proudly in front of the earth station in Andover. People cheered and whooped and wept. A new era in communication had begun.

Telstar ushered in the era of satellite communications. Communicating by satellite is a simple concept: Messages are sent up from one place on Earth and "bounced" off the satellite to another, far distant spot on Earth. But how does the satellite stay in the sky? How does it get there in the first place? How does it "bounce" the message to just the right place? What kind of messages are they? The answers are even more fascinating than the questions.

HOW SATELLITES STAY IN ORBIT

A satellite does not have to be a machine made by humans, launched into space, and placed into a certain orbit. A satellite is any object that circles around a larger body. Earth and all the other planets are satellites of the Sun. A satellite can have its own satellites; the Moon, for example, is a satellite of Earth. *Telstar* and the hundreds of communications satellites that were built after its successful debut are artificial satellites of Earth.

An Earth satellite can revolve around the globe because of two factors: inertia and gravity. Inertia is the tendency of any physical object to either stay at rest or to keep moving in a certain direction and at a certain speed until an outside force interferes with the object's speed and direction. A satellite, whether natural or artificial, is moving all the time. It moves in one direction away from Earth. There is virtually no object in space to make it stop moving or alter its direction. What keeps it from flying way beyond Earth into deepest space is gravity. Gravity pulls it toward Earth at the same time inertia propels it away from Earth. When these two factors are balanced exactly, the satellite keeps moving but stays "tied" to Earth; it orbits Earth on a regular and predictable path.

For a satellite to achieve the perfect balance between inertia and gravity, the satellite has to move at just the right speed. If it moves too fast, the force of motion will propel it beyond the power of gravity to hold it in a stable orbit. If the satellite moves too slowly, gravity pulls it down out of orbit, and it crashes into Earth. The correct speed—the speed that allows it to stay securely in orbit—is called orbital velocity. The exact speed necessary to keep a satellite orbiting the planet depends on how far the satellite is from Earth.

Satellites orbit Earth at different distances, or altitudes. The lower the altitude, the stronger the gravitational pull is

Syncom IV just after being launched from the space shuttle Discovery in April 1985. It was supposed to be placed into geosynchronous orbit. The satellite's booster did not fire correctly and *Syncom IV* failed to reach the proper orbit. It was repaired on a later shuttle mission.

because Earth's mass and atmosphere exert a powerful pull. The stronger the gravitational pull, the faster the satellite must go to avoid being drawn into Earth. Therefore, the closer a satellite is to Earth, the higher its speed, or its orbital velocity, must be. Objects far from Earth, where the force of gravity on them is weaker, do not need as high an orbital velocity. The Moon is 240,000 miles (386,243 km) above Earth. It has an orbital velocity of 2,300 mph (3,701.5 km/hr). A satellite in an orbit more than ten times lower, 22,300 miles (35,888 km) above Earth, requires an orbital velocity of about 7,000 mph (11,265 km/hr). A low-orbit satellite, one just a few hundred miles up, must orbit at about 17,000 mph (27,359 km/hr), which is almost five miles (8 km) a second!

AROUND THE WORLD

In addition to determining a satellite's orbital velocity, altitude also determines how long it takes the satellite to circle Earth. The time it takes to make one revolution around Earth is called the satellite's period. The closer a satellite is to Earth, the less distance it must cover to circle the planet. So low-orbit satellites have shorter periods than high-orbit satellites; they get around Earth faster. The Moon has a period of 27.3 days. A low-orbit satellite can have a period as brief as ninety minutes. With a period of this length, the satellite would pass over the same spot on Earth sixteen times a day.

Long ago, scientists discovered that a satellite orbiting at 22,300 miles (35,888 km) would have a period of exactly twenty-four hours. Instead of passing over a place on Earth several times a day, it would circle Earth at the same rate Earth is rotating, remaining over a fixed spot below. If its orbit were around the equator, it would look to an observer below like it was not moving at all. Such a satellite is called geosynchronous (meaning it is in synch with Earth's movement) or geostationary (meaning it appears to float motionless in the sky).

LAUNCHING A SATELLITE

Placing a satellite into orbit is a two-step process. First, the satellite has to be taken to the right altitude. Then it has to be pushed off at the right velocity to begin and maintain its orbit. A satellite cannot perform either of these actions by itself.

The satellite must be brought to its proper altitude by a launch vehicle, such as a rocket or a space shuttle. The most difficult part of the rocket's journey is the beginning. The atmosphere just above Earth is much thicker than the atmosphere in space, and the rocket must blast through that thickness. In addition, the rocket itself is heaviest at the beginning of its journey because it has a full load of fuel. So it is aimed straight up, taking the shortest route to the desired altitude.

A satellite launch vehicle is a multistage rocket; in addition to the main rocket, it also contains several smaller booster rockets. The boosters give the rocket its initial thrust during launch and carry the vehicle through the early part of the flight. As the fuel in the boosters is used up, the rockets fall away and other boosters take over. Each time a booster is dropped, the vehicle gets lighter, so it can travel faster.

Once the launch vehicle is in the thinner air of outer space, about 120 miles (193 km) up, a guidance system

A rocket carrying a communications satellite into orbit pierces the cloudy sky above Cape Canaveral, Florida. The rocket will release the satellite when it reaches the proper altitude and speed, as determined by space technicians.

within the main rocket is used to tilt it eastward. Earth rotates to the east, and a rocket heading in the same direction gets an extra push and a little more speed from the force of Earth's rotation. When it finally reaches the proper altitude and speed, it releases its cargo and the satellite flies free of the spent rocket.

The satellite is usually not in perfect orbit right away. It has small rockets attached to its body, however, that allow technicians on the ground to correct its path and velocity. By firing the right rockets at the right time, the technicians are able to place the satellite in the correct orbit for receiving and sending messages to and from Earth.

COMMUNICATING BY SATELLITE

The messages that the satellite sends back to Earth—radio broadcasts, television signals, phone conversations, and other information—are translated into electrical signals. Electrical signals travel in an up-and-down motion—a wave—and the number of times they go up and down in a second is called frequency. The electrical signals that make up different types of messages have different frequencies. So, for example, a television broadcast's frequency is different from that of a radio broadcast, which is different from that of a cellular phone call.

Satellite dishes on Earth's surface serve three purposes: to send commands to a satellite in orbit, to transmit information that the satellite will relay to other satellites and ground stations throughout the world, and to receive information from satellites, such as television broadcasts and cell phone calls.

Communicating these messages by satellite involves:

- An uplink, which sends a message from Earth up to the satellite. This requires a ground station that can send signals and a satellite equipped with a receiver and a receive antenna.
- A downlink, which sends a message from the satellite down to Earth. This requires a satellite equipped with a transmitter and a transmission antenna on the satellite and a receiving station on the ground with a receiver and a receive antenna.

TRANSCRIPT OF FIRST LIVE TELEPHONE CONVERSATION TRANSMITTED BY SATELLITE, *TELSTAR*, JULY 10, 1962

Frederick Kappel, AT&T board chairman: *"Good evening, Mr. Vice President. This is Fred Kappel calling from the earth station at Andover, Maine. The call is being relayed through our* Telstar *satellite, as I'm sure you know. How do you hear me?"*

Vice President Lyndon Johnson: *"You're coming through nicely, Mr. Kappel.*

Frederick Kappel: *"Well, that's wonderful."*
— From AT&T Web site (http://www.att.com/technology/history)

- A connection on the satellite between the uplink and the downlink.
- A supply of power that keeps all satellite systems operating.

In the uplink, a transmitter on the ground takes the information to be sent and encodes it into a radio wave of a specific frequency. The transmitter antenna, which is usually a round dish, sends the signal into the air toward the satellite. The receive antenna on the satellite captures the signal. The receiver on the satellite decodes the message because it will be sent back to Earth at a different frequency. Different frequencies are used to send and receive signals. If an incoming and outgoing signal were on the exact same frequency, they might interfere with or scramble each other.

The downlink begins when the satellite transmitter encodes the message sent from one ground station into a radio wave of the proper frequency. The satellite transmit antenna then sends the message to a different ground station on Earth. The receiving antenna at the ground station captures the message, and the receiver decodes it into television pictures, radio programs, telephone conversations, faxes, or Internet messages, as the case may be. Some satellites broadcast messages far and wide, whereas others direct the message by narrow beams to very specific spots.

POWERING THE SATELLITE

All of this requires a source of energy. The only source of power available hundreds or thousands of miles out in space is the Sun. So a communications satellite uses solar energy to power its operations. It features panels covered with solar cells that must collect and provide enough power for the satellite's transmitters and receivers, a computer that monitors and controls all the satellite's systems, and

A technician examines a satellite's solar panel. Solar panels help to power satellites by generating energy from sunlight. Damage to the solar panels can cause a satellite to malfunction or fail to orbit properly.

batteries that store extra power for those periods when Earth passes between the satellite and the Sun, blocking the Sun's rays.

The solar panels generate only a few thousand watts of power, but that is enough to broadcast messages all around the globe. Actually, one satellite cannot send a signal all across the planet. Usually two or more satellites work together as a sort of relay team. A single satellite can receive and transmit signals only in a limited area, but a system of several satellites can have a broader scope. One satellite in geosynchronous orbit can "see" about a third of Earth. Therefore, a system of three satellites at equal distances from one another can see the whole planet at once. A signal from one ground station can be bounced from one satellite to the next and sent to receiver antennas at ground stations located anywhere on Earth.

CHAPTER TWO

THE GLOBAL VILLAGE

As exciting as the 1962 *Telstar* launch was, it was not the first satellite to be launched into space. It was not even the first communications satellite. It was the first "real-time" communications satellite—the first vehicle of any kind that permitted voices and images originating on one continent to be broadcast in another at almost precisely the same instant. This allowed for live coverage of unfolding events—such as news stories and sporting events. It was the first step in creating what some were calling the global village—a world in which all people were neighbors, in close communication and discussing shared information. No one was farther away from anyone else, anywhere in the world, than a simple telephone call or television antenna.

Even before *Telstar*, communication between distant points had been getting faster and faster, so that the world seemed, in a sense, to be getting smaller. In the early centuries of human history, long-distance messages had

In addition to inventing the code that bears his name, Samuel Morse (1791–1872) aggressively promoted the telegraph until it achieved widespread use across the United States in the nineteenth century.

been carried by physical means: in person; through the use of drums, sirens, smoke signals, bells, and horns; or by carrier pigeon, horseback, or ships. The first important breakthrough in speeding up communication between far-flung locales was the shift from the use of physical methods of moving messages to electrical means.

ELECTRICAL COMMUNICATIONS

In 1837, William Cooke and Charles Wheatstone in England and Samuel Morse in the United States discovered a way to use electricity to send messages. Morse devised an alphabet of dots and dashes—short and long electrical pulses—known as Morse code. Telegraph operators could spell words with these impulses and send them through wires over a distance of twenty miles (32.2 km). Morse worked at his invention until he devised relays, allowing a series of twenty mile stretches to be stitched together, creating a longer range of communication.

Soon after the telegraph was introduced, companies and governments placed the wires and relays in cables and laid those cables throughout much of the world. In the early 1850s, telegraph lines connected cities in Europe, cities in North America, and cities in the Middle East. In 1866, two cables were laid across the Atlantic Ocean, joining the United States and England. Before the century ended, telegraph lines connected six continents, and the world seemed knit together tightly by telegraph wires humming with communication. The world did indeed seem smaller, more immediate, more intimate.

Telegraph transmission was slow, however. At first, only one message could be sent across a wire at a time, and that message had to be spelled out, letter by letter. Electricians looked for ways to transmit more signals on each wire. Thomas Edison's quadruplex, a variation of the telegraph, allowed four messages to be sent over the same wire simultaneously—two in one direction and two in the other. An English telegraph of 1883 allowed larger numbers of words to be transmitted over a wire at once. Eventually, inventors began experimenting with sending different tones over the wires instead of simple electrical pulses. It was then a small leap from musical tones to the transmission of the human voice. In 1876,

Alexander Graham Bell, hoping to improve the telegraph, instead invented the telephone.

It would take thirty-five years to develop the technology to a point where telephone messages could be transmitted more than 100 miles (161 km). In 1911, one line operated between New York and Denver, and in 1915, the American Telegraph and Telephone Company (AT&T) strung a transcontinental telephone line from New York City to San Francisco. The first cable to connect two continents by telephone was laid at the bottom of the Atlantic Ocean in 1956, and the world grew smaller still. Communication was still limited, however, to locations where wires could be placed. Many remote and rugged areas remained unconnected.

WIRELESS COMMUNICATION

That limitation was removed in 1896. Physicists had discovered electromagnetic waves, and the Italian inventor Guglielmo Marconi devised a method of sending messages over those waves. These messages could travel through the air several hundred miles without wires. Instead of cables, this wireless telegraphy used transmitters and receivers. A transmitter antenna sent the message, and a receiver antenna "caught" it. This first form of wireless communication was known as radio.

Guglielmo Marconi, who invented the wireless telegraph, was awarded the Nobel Prize for Physics in 1909. The company he formed in 1897, the Wireless and Signal Company, survives today as Marconi, Plc.

At first, radios were used primarily to communicate between ships at sea and from ship to shore. During World War I, they were used to send messages between battlefield and command stations. The use of radio to broadcast entertainment occurred almost as an accident. Some people who understood the technology built their own radio sets and became amateur radio operators, tuning in to whatever signals they could catch around them.

One such amateur, Frank Conrad, was an engineer with the Westinghouse Electric Company (founded by George Westinghouse Jr., an inventor whose work helped revolutionize the railroad, telegraph, and telephone industries and usher in the era of electricity). From his radio, Conrad broadcast music, and other amateur operators near him picked up the sounds. Conrad saw a business opportunity and, in 1920, founded the first

Frank Conrad broadcasts from his little radio set in Wilkensburg, Pennsylvania, in 1924. Despite his modest seventh-grade education, he was awarded an honorary doctor of science degree from the University of Pittsburgh for his contributions to the development of electronics.

commercial radio station, KDKA, in Pittsburgh (which is still in operation). Before long, towers that could relay radio signals from one antenna to the next dotted the countryside.

By 1945, the commercial use of radio had brought the world even closer together. Speeches and concerts on the East Coast could be heard live in Oregon and California. During World War II, news from London was broadcast regularly in the United States. More remote places outside the reach of antenna towers could not receive radio transmissions, however.

Nor could they receive the newest form of news and entertainment that was just beginning to captivate the public imagination—television. Television signals were more complex, and they could be carried only about 100 miles (161 km). If the latest forms of communication were to be truly global, innovative technology would be needed to carry these new broadcast signals.

SCIENCE FICTION BECOMES SCIENCE FACT

The first idea for how to develop the new technology necessary for global broadcast of television signals came from an unlikely source. English scientist and writer Arthur C. Clarke, who wrote many science fiction stories, including the one the film *2001: A Space Odyssey* was based on, also wrote serious articles explaining how radio waves, microwaves, transmitters, receivers, and amplifiers worked.

Hoping to show how communications could be improved, he described his theory in a 1945 article entitled "Extra-Terrestrial Relays" in the magazine *Wireless World*. Clarke suggested building three artificial satellites that would circle Earth much like the Moon does. They would serve as relay stations: Radio waves would be sent from a ground station at one location on Earth, bounced from one satellite to another, and sent back to a ground station at a different location. Each of the three satellites could send and receive signals to and from one-third of Earth's surface. Taken together, their range would extend over the entire surface of Earth. The three satellites, therefore, could provide almost instant

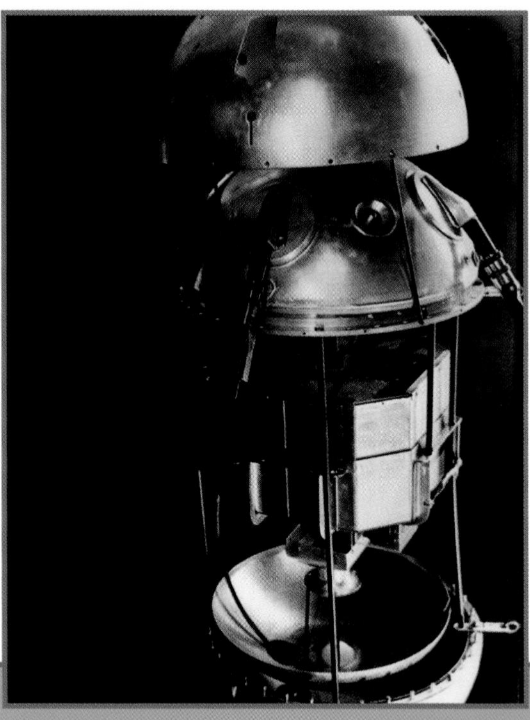

The successful launch of *Sputnik* I by the Soviet Union in 1957 triggered an intense focus on the education and development of scientists and engineers in the United States.

communication between any two spots on the globe. The logic appeared to be sound, but in 1945, Clarke's article struck many readers as more like one of his far-fetched science fiction stories than serious scientific research

Twelve years later, however, when the Soviet Union astonished the world by firing the first satellite, *Sputnik I*, into orbit, Clarke's idea no longer seemed so outlandish. The United States scrambled to keep up with the Soviets and built a satellite of its own.

Less than four months after *Sputnik I* stunned the world, the United States launched its first satellite, named *Explorer I*. *Explorer* was a scientific satellite, a research tool that gathered information about space. In the frantic rush to do more and better than the Soviet Union, a number of other space exploration satellites followed. Once the race to match and surpass the Soviets was won, scientists began to turn their attention to the possibility of using satellites to improve communications.

In 1958, one year after *Sputnik*, the United States created the National Aeronautics and Space Administration (NASA). Its main purpose was to research the possibility and challenges of flight within and outside Earth's atmosphere. Conquering space was considered a matter of national security. NASA was a government agency, but it was not connected to the military. Therefore, it was given the assignment of developing low-orbit communications satellites, both passive and active.

PASSIVE COMMUNICATIONS SATELLITES

A passive communications satellite simply reflects radio waves. It has no equipment for doing anything with the waves other than bouncing them off its surface. It orbits relatively low in the sky, and messages are bounced off it from one spot on Earth to another many hundreds of miles away. It is also called a reflector satellite.

As early as 1954, the U.S. Navy had been trying to use the Moon as a passive satellite. The experiment worked, and the first functioning communications system in space was created. It was called Communication by Moon Relay (CMR). From 1959 to 1963, naval technicians sent messages between Washington, D.C., and Hawaii by bouncing radio waves off Earth's natural satellite—the moon.

TRANSCRIPT OF 1960 TRANSMISSION FROM *ECHO*

"This is President Eisenhower speaking. It is a great personal satisfaction to participate in this first experiment in communications involving the use of a satellite balloon known as Echo.*"*

—From AT&T Web site (http://www.att.com/technology/history)

NASA's first attempt to build an artificial communications satellite was also a success. It was called *Echo* because it was expected to reflect radio and television signals in much the same way that canyon walls can echo human voices. *Echo* was basically a huge balloon, 100 feet (30.5 meters) across, made of polyester and coated with aluminum—something like a giant Mylar balloon. It was rocketed into orbit and then inflated. Signals were sent from a transmitting station and reflected off the satellite's smooth, shiny surface. The signals could then be picked up, or received, by any antenna that could "see" the satellite.

The only problem was that no antenna could see the satellite for an extended period of time. *Echo* orbited Earth once every ninety minutes, and it stayed in the range of any given earth station for only about ten of those ninety minutes. Nevertheless, *Echo* showed that satellite communication would work and was far better than the alternative of sending broadcast signals along a complicated chain of ground relay stations. As exciting as *Echo*'s success was, however, an even more promising possibility was taking shape: the active communications satellite.

ACTIVE COMMUNICATIONS SATELLITES

Almost two years before *Echo* was launched, at the end of 1958, the U.S. Army introduced the first active communications satellite, called *SCORE*. The letters in the name stood for "Signal Communication by Orbiting Relay Equipment." The name meant that radio or television signals were relayed by equipment that orbited Earth. The equipment did more than reflect the signals back to Earth. It amplified them, making them stronger. This meant that the quality of the communication was better than that of a passive, or reflector, satellite.

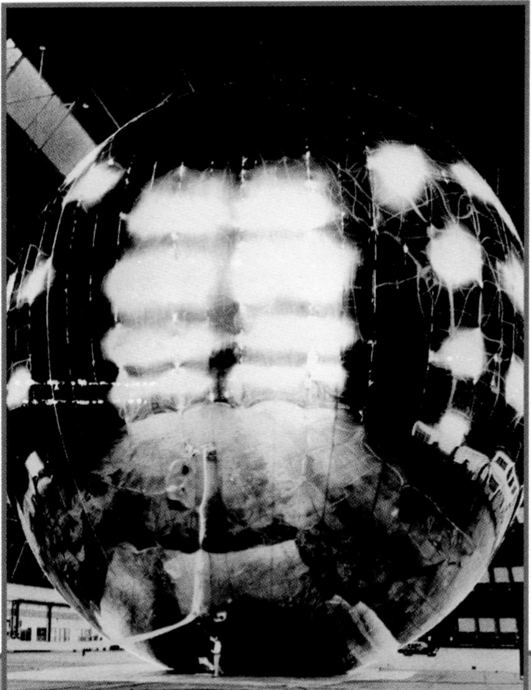

The *Echo I* satellite was successfully placed into orbit in 1960. It was used to redirect radio and television signals across the United States and was a step toward eventual live national broadcasts.

But *SCORE* had the same problem as *Echo*: It could transmit to or receive signals from a given ground station for only about ten minutes at a time. The satellite received a message as it passed over one station, stored the message on a tape recorder, and retransmitted the recorded message as it passed over another station. SCORE was a limited success. Its batteries worked for only thirteen days. Two years later, the Army Signal Corp sent a similar satellite, *Courier*, into the sky; it functioned for seventeen days.

All during this time, the private company AT&T—which had started as a telegraph company—was hard at work developing its own satellite. AT&T was in the communications business, and the millions of dollars it would take to research and develop new technology would be well spent if it meant the company could provide international telephone service better, faster, sooner, and—eventually—more cheaply than its competitors. At its Bell Laboratories, AT&T engineers began in 1960 to develop a communications system called Telstar. The plan was to use 50 to 120 satellites and two dozen ground stations to reach every point on the globe. By 1962, Bell Labs had built six experimental satellites, and NASA had agreed to launch the first two for a fee.

The Telstar program was an overwhelming success. The embarrassment Americans felt for not being the first to reach space was forgotten. *Sputnik* had stayed up for only twenty-one days and did nothing but beep, but Telstar provided real-time communication across continents. Telstar proved that global communication was possible, and, AT&T hoped, very profitable.

CHAPTER THREE

THE BUSINESS OF SATELLITE COMMUNICATION

Telstar revolutionized the communications industry, demonstrating that satellites could provide broader and clearer communication than any Earth-based systems could (such as phone lines and a vast series of ground-based relay stations). It looked like satellites could make money for communications companies with deep financial resources, such as AT&T. Yet a number of smaller businesses also wanted to develop their own satellites. They found, however, that building and launching them was too expensive. They needed to find a partner to help them get their machines into space.

RELAY AND SYNCOM

One organization that had money to spend on research and construction of satellites was NASA. After AT&T began developing the Telstar system with its own money, NASA decided to work in cooperation with private corporations in the development of satellites. The corporation would

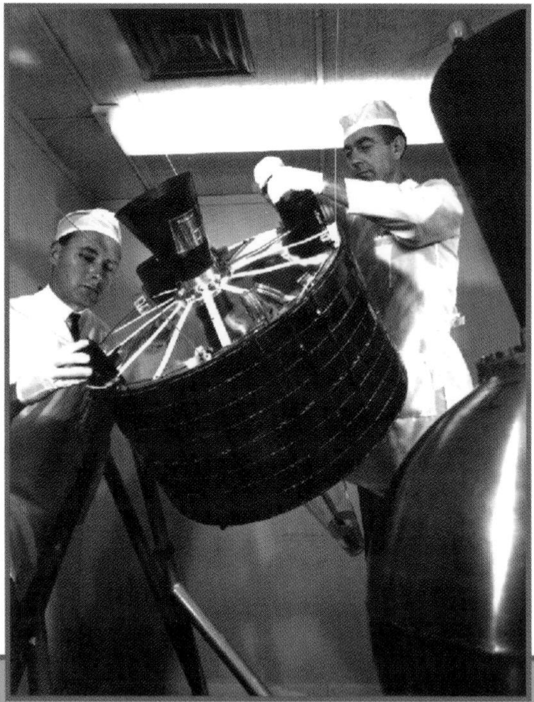

Technicians work on *Syncom 1* to prepare it for its launch on February 14, 1963. *Syncom I, II*, and *III* were experimental communications satellites that transmitted voice, teletype, and facsimile signals.

design and build the satellite, and NASA would pay for the work and the launch. NASA did not choose to work with AT&T but instead selected its competitors. In the early experimental stages, NASA awarded a contract for a medium-orbit satellite to Radio Corporation of America (RCA) and a contract for a high-orbit, geosynchronous satellite to Hughes Aircraft Company (now Boeing).

RCA built and launched its satellite, called *Relay*, very soon after *Telstar* electrified the world. *Relay* was very much like *Telstar*, so its launch received very little attention. The Hughes satellite, however, called *Syncom*, was an entirely new kind of satellite. Launched only one year after *Telstar*, it made the technology in *Telstar* and *Relay* obsolete overnight.

Syncom was the first test of Arthur C. Clarke's theory of "extra-terrestrial relays," presented nearly twenty years earlier. In 1963, the technology existed to get a satellite 22,300 miles (35,888 km) above Earth. Would three such satellites be able to blanket Earth with television coverage? Hughes Aircraft Company hoped so. With its $4 million contract from NASA, Hughes built three Syncoms.

The first one worked for only a few minutes; it failed when technicians tried to correct its orbit. Hughes was not discouraged, however. Learning from the error, the company made three changes to the model and sent *Syncom 2* into the sky five months later. This second geosynchronous satellite worked! It provided high-quality, instant communication across the Atlantic Ocean, a third of the world away.

Hughes rushed to get its third experimental model—and second functioning satellite—into orbit. Its goal was to use its two satellites to broadcast not only across the Atlantic, but also across the Pacific. Just one month before the 1964 Olympics in Tokyo, Japan, *Syncom 3* was launched. Audiences in two-thirds of the world watched parts of the games live, as they were happening. Satellite communication was on its way to becoming a very lucrative use of space.

REGULATING THE BUSINESS

Which company would be the one to make a fortune with satellites? In 1961, the answer seemed very likely to be

AT&T. It was the biggest communications company in the United States and the only company with the equipment necessary for international communications. It had enough money to build and test experimental satellites. It had a state-of-the-art laboratory and world-class scientists and engineers. And it wanted to be the one.

U.S. president John F. Kennedy, however, did not want any one company to control the new technology. AT&T already had a monopoly on land communications, and Kennedy felt it would be unfair for the company to have a monopoly on space communications as well. Congress held hearings to determine who would be allowed to create and use satellites for commercial (for-profit) purposes.

Some people wanted the government to control space communications, arguing that the airwaves were public. Private companies, such as AT&T, RCA, and Hughes, pointed out that they could develop communications that would benefit everyone far more quickly and efficiently than the government. Finally, a compromise was reached. In 1962, less than two months after *Telstar* was launched, President Kennedy signed the Communications Satellite Act, which created a new company, Comsat.

COMSAT

Comsat, the Communications Satellite Corporation, was a government corporation that would operate through

A Comsat earth station in Andover, Maine, in 1962. AT&T, the maker of *Telstar*, had hopes to dominate the emerging satellite communications business, as it had done with the telephone industry. By forming Comsat, President John F. Kennedy ensured that the satellite communications industry would be open to competition but regulated and controlled by the government.

private businesses. Comsat would use private companies to design and produce satellite systems, but it would own the systems once they were constructed and operational. This way, the satellites would remain publicly owned. Comsat could not sell satellite services directly to the customers who would use them, however. It could sell those services only to the communications companies, who would in turn sell them to the public, insuring that private enterprise and the government could exercise control. This arrangement was a quasi-government corporation—part government, or public, and part private. Comsat was the only company in the United States permitted to own and operate communications satellites.

INTELSAT

As more and more countries began to make their own satellites, some new questions arose. Why should one country spend millions of dollars to build and launch a satellite when it could save money by simply renting the services of one already in the sky? What would happen if two countries tried to put a satellite in the same position in space? Would signals from one country interfere with those from another? What could one country do if it was picking up unwanted messages from another country's satellite? What if a country did not want its messages being intercepted by a foreign satellite? The need for international cooperation and regulation was obvious.

So, in 1964, representatives of eighteen nations formed a worldwide satellite network, now called the International Telecommunications Satellite Organization, or Intelsat, that would provide satellite communication services to paying customers the world over.

Immediately after it was formed, Intelsat set about to create the world's first commercial communications satellite whose services would be available for a fee to several countries. The Intelsat satellite, first called *Early Bird* and later renamed *Intelsat 1*, was made by Comsat, the United States's representative in Intelsat. Comsat arranged for Hughes Aircraft to build the satellite and contracted with NASA to launch it. It would transmit to earth stations in four countries: the United States, England, France, and Germany. On April 6, 1965, a Delta rocket took *Early Bird* into orbit 22,300 miles (35,888 km) above Earth.

Early Bird was launched from Cape Canaveral on April 6, 1965, to relay radio, television, teletype, and telephone messages between North America and Europe.

This first international commercial communications satellite was very simple compared with today's models. It did not have a battery for energy storage, so it could operate only when it could "see" the Sun. It could transmit to only one point at a time, so the same messages could not be received in all four countries simultaneously. Still, it was a commercial success. It had

COMMUNICATIONS SERVICES

What types of communications do satellites provide? Almost any long-distance communication imaginable! Because of satellites, people can:

- **Talk on the telephone to someone thousands of miles away and hear clearly.**
- **Hear radio programs from another continent without static.**
- **Use a cell phone or a pager.**
- **Make a phone call from an airplane while it is in flight.**
- **Get on the Internet quickly.**
- **Send an e-mail message to anyone with an e-mail account, anywhere in the world.**
- **Send a fax thousands of miles away that will be received within seconds.**
- **Hold a video conference with people in several different countries.**
- **Withdraw money from an ATM in one state that was deposited in a bank in a different state.**
- **Talk from a control tower to a pilot in an airplane.**

been designed to work for only eighteen months, but it operated nonstop for almost four years. *Intelsat 1* paved the way for other commercial satellites, which eventually provided communications coverage for the entire globe.

Today, Intelsat includes over 100 member nations and offers satellite communications services to customers in more than 200 countries and territories—voice and data communications, corporate networks, Internet access, and broadcast services. It has twenty-one satellites and hundreds of ground stations. Its customers include the world's leading telecommunications carriers, Internet service providers, and broadcasting companies. In 2000, the company generated $1.1 billion in revenue. Just as AT&T had anticipated almost forty years earlier, satellite communications has indeed become a highly profitable business.

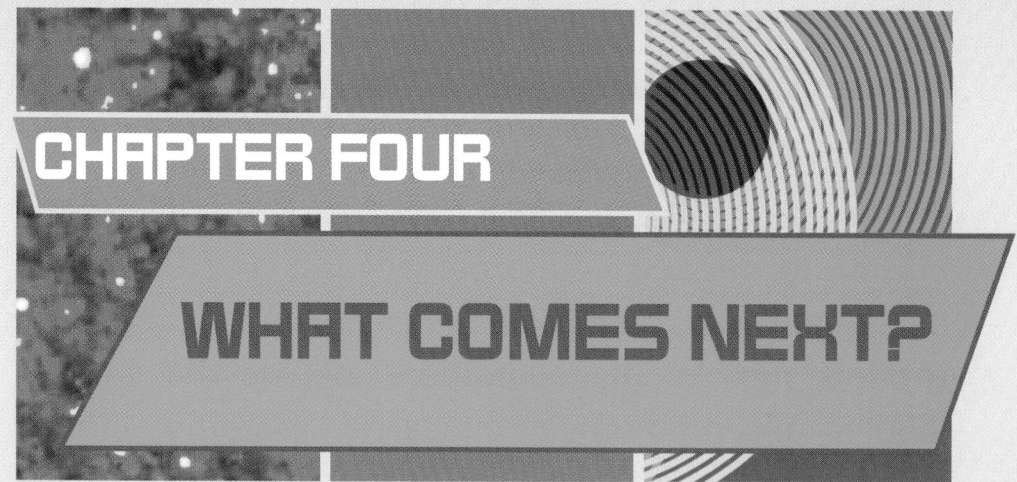

In the forty years since *Telstar*, the communications industry has changed dramatically. In the 1960s, earth stations had 100-foot (30.5 m) dish antennas that cost up to $10 million. Today's ground stations have 15-foot (4.6 m) dishes that cost $30,000 or less. Private homes have small dish antennas that are only one foot (30.5 cm) across and cost from $100 to $200. In the 1960s, the cost of a telephone call relayed by satellite was more than $10 per minute; today it is less than $1.00. In those early years, access was limited to the United States and parts of Europe. Today, satellite coverage reaches into every corner of the planet. The first satellites could transmit to only one location at a time; modern satellites can broadcast to many points at once. Early satellites could handle only small amounts of information. Digital technology has enabled today's satellites to receive and send vast quantities of complex data. Today, more information is available to more people at a faster speed and a lower cost.

Where will the industry go from here? It will continue to develop new types of satellites that will allow it to offer even more services to even more customers at an even better price.

LEOS

One of the newest ideas in satellite technology is actually a very old one. Back in the 1960s, high-altitude geostationary satellites replaced low- and medium-orbit satellites for communications. Now most companies are abandoning the high-orbit spacecraft and going back to satellites in low-earth orbit (LEO).

One of the reasons the low- and medium-orbit satellites were discontinued was that the early models experienced electrical problems when they traveled through belts of radiation that surround the lower reaches of Earth's atmosphere. Now, improved technology has allowed scientists to program the lower satellites' orbits so that they miss the radiation belts. With this obstacle to smooth functioning out of the way, communications companies are focusing on the advantages of lower-orbiting satellites.

One advantage of LEOs is their ability to specialize. A single geostationary satellite processes many different types of signals: voice, radio, television, data, and so on. A LEO can focus all its systems on one task, such as providing cellular phone service. Companies that provide only one type of service can use a LEO that receives and transmits only that type of signal; it would make no sense for them to rent the services of an expensive satellite that also offers television, Internet, and radio transmissions.

A LEO is much less expensive than a high-orbit satellite. A geostationary satellite is placed in such a high orbit that the signals it receives and transmits must travel 22,300 miles (35,888 km) and back again. This requires more time and power than sending those same signals a few hundred miles. So a LEO can deliver the same message faster and with less energy—and therefore less expensively.

A LEO is usually smaller than a geostationary satellite, so it costs less to build. Because it is placed so low in the sky, it is also less expensive to launch. Several low-orbit satellites can be launched at the same time from the same rocket, one right after the other, reducing the costs of placing satellites in orbit. So a group of twenty or thirty LEOs can cost less to build, launch, and operate than a system of three geostationary satellites.

Such a group might also provide more complete coverage of Earth's surface. A high-orbiting satellite, in order to remain geostationary, must circle the globe at the equator. Therefore, coverage is good near the equator and poorer close to the North and South Poles. A LEO can be placed in any orbit, even one that goes over the poles. So a LEO system can provide high-quality

A Delta rocket carrying the first five Iridium satellites lifts off from the Vandenberg Air Force Base in California on February 11, 2002. Motorola, a communications and electronics company, uses a network of more than sixty low-earth orbiting Iridium satellites to provide essential communications services to virtually every corner of the world.

communications anywhere, from the extreme north of Norway to the southern tip of Argentina.

The geostationary satellite had been the satellite of choice for so long because of its very big "footprint." That is, it can "see" a large area of Earth. A big footprint brings a big workload, however; one satellite must serve a very wide area. It must handle a huge number of uplinks and downlinks. Within a group of LEOs, the same amount of work can be delegated to several satellites. As a LEO circles Earth, it passes in and out of range of a ground station quickly. But as soon as one LEO passes out of range, the next one in the system comes into view. The signals are relayed from one satellite to another. Thus a LEO system can accomplish the same things as geostationary satellites for less money. This allows communications companies to offer their services at lower prices while still providing them with a larger profit.

"STEERABLE" SATELLITES

Even though LEOs appear to be the wave of the future, that does not mean that geosynchronous satellites have become relics of the past. New technology has overcome some of the disadvantages of the high-orbit spacecraft. The latest geosynchronous satellites do not have a single downlink that broadcasts to a third of Earth at once. Instead, they have many uplink and downlink beams. They

The Advanced Communications Technology Satellite was developed as an experimental communications satellite that would bring together representatives of business, government, and academia to conduct a wide range of investigations into space-based telecommunications technology that are expected to be used widely in the twenty-first century.

can steer each beam so it focuses on a spot about 100 miles (161 km) wide. In this way, no signal strength is wasted by being diffused over a third of Earth. The beams can be moved from spot to spot in milliseconds.

NASA is experimenting with a steerable satellite it calls ACTS, or Advanced Communications Technology Satellite. NASA calls the spacecraft a "switchboard in the sky" because it functions something like an old-fashioned telephone switchboard: An operator determines who wants to connect with whom and plugs them in. The uplink and downlink beams are then steered, or pointed, to the customer's earth station and the connection is made.

TARGETED SERVICES

The geostationary satellite is an all-purpose satellite used for many different types of communication, such as television and radio broadcasts, long-distance phone service, and Internet access. LEOs target a particular service. For example, one company may use a low-orbit communications satellite for nothing but wireless e-mail. About half of all communications companies limit their satellites to direct home reception of television programs. Teledesic, an information technology company, is planning to launch thirty medium-orbit satellites to provide only one service, "Internet-in-the-sky," a satellite-based Internet provider.

Motorola is producing an even bigger LEO system, a network of sixty-six low-orbit satellites. Called Iridium, it will be used strictly for wireless telephone service. By the time all sixty-six satellites are in orbit, the phone service will be available from any place on the globe, and the project will have cost a minimum of $5 billion.

At least three companies are building satellites for radio service only. The radios that use these systems need a receive antenna and a receiver, just as an earth station does. An antenna only two or three inches long (5.1 to 7.6

cm) can receive a signal on Earth directly from the satellite. Radio-only satellites will enable people to tune their home or car radios to more than 100 stations scattered thousands of miles apart. Local radio stations will no longer be the only choices available. The stations will not fade as cars travel because the satellites will have nearly global footprints.

SATELLITES VERSUS OTHER CONNECTION MODES

All of these services—television, e-mail, cellular telephone, radio, and the Internet—are currently available without the use of satellites. They are accessible through repeaters, such as towers, that relay signals across distances and through overhead or underground cables. Satellite communications is a business like any other, and its future depends on whether it can provide these communication services cheaper than today's land-based providers can.

When satellites began relaying telephone conversations, they were much cheaper than the standard system of cable phone lines. Satellites could handle ten times as many calls as transoceanic cables for about a tenth of the price. Once fiber-optic cables were developed, however, the cost of cable dropped sharply. Fiber optics, which allow for the rapid flow of a tremendous volume of information, may be able to provide communications for

Although the development of fiber-optic cables has lowered the cost of services provided by land-based communication networks, satellites are still a cheaper option for long distance communication.

about the same price as satellites can.

Satellites, however, have two advantages over cable. First, with cable, the price increases with distance: The longer the message's journey, the more it costs. A long-distance phone call carried by fiber-optic cable from New York to California will cost more than one from New York to Boston. With satellites, the distance between the caller and the receiver of the call does not affect cost. Two points on Earth that are far from each other are about the same distance from the satellite that is connecting the two points. And with satellites, there is no additional expense of laying and maintaining the cable, a cost that gets passed on to consumers. So for very long-distance communication, satellites may still be cheaper than fiber optics.

Second, any cable-based system can provide only point-to-point connection in which one location—a caller,

JAMES EARLY, DEVELOPER OF SOLAR CELLS AND TRANSISTORS FOR *TELSTAR*

"The success of [**Telstar**] told us unmistakably that all of humanity were members of a single global community...Mankind was tied together as never before.** Telstar **worked. Real-time television had leaped the Atlantic. My eyes watered."

—From Southwest Museum of Engineering, Communications, and Computation Web site (http://www.smecc.org)

a computer—is connected to only one other location. Satellites can provide point-to-multi-point communication in which numerous locations can be connected simultaneously and linked to a central site. What one satellite can accomplish could require several expensive and long-distance fiber-optic cables.

The demand for long-distance communication will only increase in the future. Very few places in the world are so remote and self-contained that they require only local phone and Internet connections and homegrown radio and television broadcasts. Business transactions, live entertainment, educational instruction, even personal relationships—much of modern life involves interaction that takes place across thousands of miles. All of it depends on clear, fast, and affordable communication. The smaller the world becomes, the more it will rely on satellites to provide the ties that bind us all together.

CONCLUSION

The world in which *Telstar* beamed the image of the American flag into thousands of living rooms on two continents was far different from the world of today. In 1962, people could choose from only three networks and a few local channels. They had to have large antennas attached to the roofs of their houses, and the antennas had to be turned in the direction of the television broadcasting station. Reception was often fuzzy, especially during storms. Not everyone had television, hardly any families had more than one, and most of them were black-and-white sets. Radios often crackled with static. Computers were huge pieces of floor-to-ceiling equipment that took up whole floors of office buildings. Long-distance phone calls had to be placed by an operator, and they were expensive. Sending mail from the United States to Europe could take a week or longer. Into this slow-moving, static-prone, inefficient world, *Telstar* was born. Instant communication had seemed like a dream, and *Telstar* made it a reality.

Before the age of satellites, communications were comparatively primitive, as can be seen in this collection of photos that includes (clockwise from bottom left) early radios (one with a ribbon dress worn by the listener that served as an antenna), unwieldy microphones used in a radio broadcast, a two-piece telephone that required the assistance of an operator to complete the connection, a large mainframe computer, and a television antenna that was difficult to mount on one's roof.

Modern communications satellites provide services that surpass what the Bell Labs engineers ever dared to imagine: hundreds of television channels that broadcast directly to three-foot (0.9 m) dishes atop private homes; crystal-clear reception of the same radio station from coast to coast; table-top, laptop, and hand-held personal computers; telephone calls placed from airplanes; e-mail and instant messaging. Communications satellites have made the 1962 dream of a global village come true.

A BRIGHTER FUTURE?

A global village is a world in which all people seem to be living close together, though they are widely scattered across the planet. When people live in close quarters, they sometimes come into conflict. The satellite technology that has drawn us together in unprecedented ways has also created some friction. When one company or one country has the ability to broadcast its signals to a wide area, some of its programming can spill over into another company's or country's territory. If these broadcasts are unwelcome, conflicts can erupt.

So far, people have been able to work out these differences. The steerable satellites and the LEOs that have narrower beams may be able to eliminate problems like these by targeting their transmissions

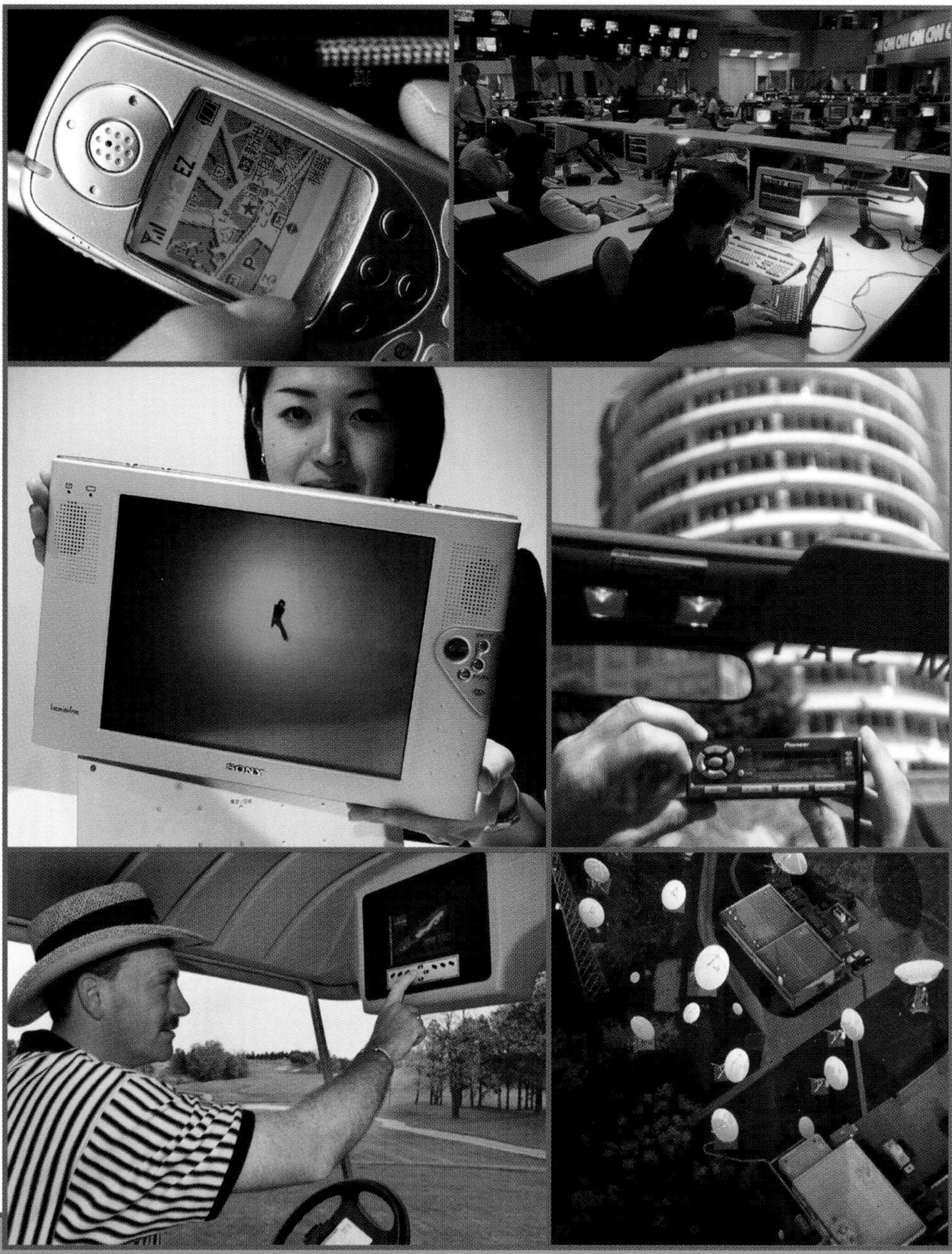

Communications satellites have allowed for such modern conveniences as (clockwise from bottom left) global positioning systems that provide users (such as hikers and drivers) with their exact locations at any given time, personal IT televisions, mobile phones, office-wide computer networks, universal tuners (which allow for long distance reception of radio stations), and satellite dishes.

more precisely. As the skies become more crowded, the airwaves busier, and competition stronger, however, other problems will surely arise. Just as conflicts in families are resolved when everyone makes an effort to understand and cooperate, so disputes in the global family will be settled when people and nations work together toward a common goal.

That is what happened with *Telstar*: The United States, Great Britain, and France collaborated to achieve the first transatlantic television broadcast. That is what is happening with Comsat: Private companies and the government are working together to deliver the benefits of satellites to the public. The same thing, too, is happening with Intelsat: More than 100 countries are accomplishing jointly what no one nation could do alone.

Satellite technology has indeed made the world seem smaller. The shared attempt to make satellite communications profitable has made it more cooperative. The results—readily available, instant connections to billions of people worldwide—will hopefully make it more united.

GLOSSARY

downlink Transmission of information from a satellite to Earth.

footprint The area of Earth that is in a satellite's range; the area it can "see."

frequency The number of times an electrical signal goes up and down in a wave pattern in one second.

geostationary Appearing to remain in the same place in relation to Earth.

geosynchronous Operating in synch with Earth's movement.

gravity The force of an object that attracts or pulls other objects toward itself.

inertia The property of a physical body to remain at rest or in motion until acted upon by an external force.

LEO A low-earth-orbit satellite.

monopoly To have sole ownership or control over a commodity or industry.

orbit The path of a satellite around its center of attraction; to revolve in such a path.

orbital velocity The speed a satellite requires to achieve a balance between its inertia and gravity, preventing it from flying off into deep space or crashing down to Earth.

period The time it takes for a satellite to complete one orbit.

real-time broadcast An event being seen or heard at the same time that it is happening.

receiver A device that "catches" and decodes messages sent through electrical signals.

repeater A device that receives and retransmits an electrical signal.

satellite An object that revolves around a larger object in a regular and predictable orbit.

telegraph A means of sending messages to distant points through electrical pulses.

transmitter Device that sends messages through electrical signals.

uplink Transmission of information from Earth to a satellite.

velocity The speed of an object as it travels through space.

Bell Laboratories
Lucent Technologies
Corporate Headquarters
600 Mountain Avenue
Murray Hill, NJ 07974
Web site: http://www.bell-labs.com

Comsat General Corporation
6550 Rock Spring Drive
Rock Spring One, 4th Floor
Bethesda, MD 20817
(301) 214-3400

Intelsat Global Service Corporation
3400 International Drive NW
Washington, DC 20008
(202) 944-6800
Web site: http://www.intelsat.com

Jet Propulsion Laboratory
Public Services Office
Mail Stop 186-113
4800 Oak Grove Drive
Pasadena, CA 91109
(818) 354-9314
Web site: http://www.jpl.nasa.gov

Kennedy Space Center
Public Inquiries
KSC, FL 32899
(321) 867-5000
Web site: http://www.ksc.nasa.gov

NASA Headquarters
Information Center
Washington, DC 20546-0001
(202) 358-0000
Web site: http://www.nasa.gov

WEB SITES

Due to the changing nature of Internet links, the Rosen Publishing Group, Inc., has developed an online list of Web sites related to the subject of this book. This site is updated regularly. Please use this link to access the list:

http://www.rosenlinks.com/ls/cosa/

FOR FURTHER READING

Branley, Franklyn Mansfield. *From Sputnik to Space Shuttles: Into the New Space Age.* New York: HarperCollins, 1989.

Branley, Franklyn Mansfield, and Sally J. Bensusen. *Mysteries of the Satellites.* New York: Dutton, 1986.

Irvine, Matthew. *Telesatellite.* New York: Franklin Watts, 1989.

MacLeod, Elizabeth. *The Phone Book: Instant Communication from Smoke Signals to Satellites and Beyond.* Toronto, ON: General Distribution Services, 1997.

Mellett, Peter, and Alex Pang. *Launching a Satellite.* Portsmouth, NH: Heinemann, 1999.

Spangenburg, Ray, and Kit Moser. *Artificial Satellites.* New York: Franklin Watts, 2001.

Brown, Gary. "How Satellites Work." Howstuffworks.com. 1998–2002. Retrieved February 2002 (http://howstuffworks.lycoszone.com/satellite.htm).

Clifton, Daniel, ed. *Twentieth Century Day by Day*. London: Dorling Kindersley, Ltd., 2000.

Dickson, Paul. *Sputnik: The Shock of the Century.* New York: Walker and Co., 2001.

Elbert, Bruce R. *Introduction to Satellite Communication.* Norwood, MA: Artech House, 1999.

Elbert, Bruce R. *The Satellite Communication Ground Segment and Earth Station Handbook.* Norwood, MA: Artech House, 2000.

Kadish, Jules E., and Thomas W. R. East. *Satellite Communications Fundamentals.* Norwood, MA: Artech House, 2000.

Luther, Arch C., and Andrew F. Inglis. *Satellite Technology: An Introduction.* Woburn, MA: Focal Press, 1997.

Roddy, Dennis. *Satellite Communications.* New York: McGraw-Hill Professional Publishing, 2001.

INDEX

ABOUT THE AUTHOR

Ann Byers is a teacher, editor, and writer. She has four grown children. She and her husband live in California. She divides her time between the San Joaquin Valley of central California and the San Francisco Bay Area.

PHOTO CREDITS

Cover © Johnson Space Center/NASA; pp. 5, 26, 51 (top right, bottom left, and bottom right) © Hulton/Archive/Getty Images, Inc.; pp. 10, 16, 20, 23, 24, 29, 51 (top left, middle left, and middle right) © Bettmann/Corbis; p. 13 © AFP/Corbis; p. 15 © Lefkowitz/Corbis; p. 18 © Richard T. Nowitz/Corbis; pp. 32, 45 © Glenn Research Center/NASA; p. 35 © G. E. Kidder-Smith/Corbis; pp. 37, 43, 53 (bottom left and middle right) © AP/Wide World Photos; p. 48 © Digital Arts/Corbis; p. 53 (top right and bottom right) © Kevin Fleming/Corbis; p. 53 (top left and middle left) © Reuters Newmedia, Inc./Corbis.

LAYOUT AND DESIGN

Tom Forget